George T. Barker

Instructions in the Preparation, Administration and Properties of Nitrous Oxide

protoxide of nitrogen, or laughing gas

George T. Barker

Instructions in the Preparation, Administration and Properties of Nitrous Oxide
protoxide of nitrogen, or laughing gas

ISBN/EAN: 9783337340346

Printed in Europe, USA, Canada, Australia, Japan

Cover: Foto ©Andreas Hilbeck / pixelio.de

More available books at **www.hansebooks.com**

INSTRUCTIONS

IN THE

PREPARATION, ADMINISTRATION, AND PROPERTIES

OF

NITROUS OXIDE,

PROTOXIDE OF NITROGEN, OR LAUGHING GAS.

BY

GEORGE T. BARKER, D. D. S.,

PROFESSOR OF THE PRINCIPLES OF DENTAL SURGERY AND THERAPEUTICS IN THE PENNSYLVANIA
COLLEGE OF DENTAL SURGERY, ETC. ETC.

PHILADELPHIA:
RUBENCAME AND STOCKTON.
1866.

PREFACE.

THAT nitrous oxide or protoxide of nitrogen is worthy to be placed prominently in the list of anæsthetics, is a fact which is daily being demonstrated by many of the most prominent surgical and dental practitioners, as it presents certain valuable characteristics, possessed by no other agent with which we are at present acquainted. But though long used as an exhilarant in the lecture-room, its extended use as an anæsthetic is of recent date, and but little, with the exception of certain fragmentary articles in the surgical and dental journals, has been written on the subject.

An excellent monograph, by Geo. J. Ziegler, M. D., on the Medical Properties and Applications of Nitrous Oxide, has been published, and from its pages I have derived many valuable suggestions to which I shall hereafter refer. This is the only work with which I am acquainted that is devoted exclusively to this subject, but as Dr. Ziegler considers its use principally as a medicine, I have therefore deemed that a work containing suggestions as to the preparation, administration, chemical, physiological, and anæsthetic properties of nitrous oxide might not be inappropriately offered to the professional public. On these subjects much remains to be learned, but it is hoped that by offering my own experience, it may stimulate investigation and research, so that nitrous oxide may eventually *stand* or *fall* on its own intrinsic merits. I commenced its use as a "doubter," and abandoned it because of failure to produce satisfactory results from want of adequate knowledge as to its proper administration; not being satisfied with my first investigations, I again commenced its use, and from an extended experience with it in dental and surgical practice, am prepared to claim for it a leading position in the list of anæsthetics.

I have endeavored, in these pages, to present the subject in a familiar and practical manner, and to induce farther attention to an anæsthetic of great value and usefulness.

G. T. B.

PHILADELPHIA, May, 1866.

CONTENTS.

(v)

LIST OF ILLUSTRATIONS.

(vii)

INSTRUCTIONS

IN THE

PREPARATION, ADMINISTRATION, AND PRO-PERTIES OF NITROUS OXIDE.

I. ON THE PREPARATION OF NITROUS OXIDE.

THIS agent was discovered by Priestley in 1776, and by him was termed *dephlogisticated nitrous air;* it is now known under the name of nitrous oxide, protoxide of nitrogen, or laughing gas. It has for its symbol in chemical nomenclature NO, as it contains equivalent proportions of nitrogen and oxygen, these elements being in chemical combination. For its equivalent or combining number 22, a specific gravity of 1.527, 100 cubic inches weighing 47.29 grains. It is a colorless gas, of a slightly sweetish taste and pleasant smell. At a pressure of 30 atmospheres in 0° or 50 atmospheres at 45° F. condenses into a clear transparent liquid. "The liquid, when exposed under the bell glass of the air-pump, is rapidly converted into a snow-like solid." ("Fowne's Chemistry.") At a temperature from between 100° to 150° below zero it crystallizes into a clear transparent body. Some recent observations upon this subject, in confirmation of the above, are thus detailed in a late number of the "Med. Times and Gazette:"—

"*Liquefaction of Laughing Gas.*—One of the most interesting objects at a recent *soirée* at the Paris Observatory consisted in the exhibition of the liquefaction of laughing gas, the protoxide of nitrogen, by M. Bianchi. This took place at zero centigrade, under a pressure of thirty atmospheres, the fluid issuing in a small jet from a strong metallic reservoir. Received in a glass tube, it retained its liquid condition by reason of the depression of temperature, produced by evaporation, so that mercury, being introduced, could be hammered like lead. Simultaneously a body in a state of ignition, plunged into the atmosphere of the liquid in which the mercury froze, burnt with a brilliant light. On pouring the protoxide into a small platinum capsule heated to redness, the liquid was found to retain all its properties, while assuming the spheroidal state, and was still able to freeze mercury contained in little glass ampullæ. Finally, the liquid protoxide became solidified under the recipient of an air-pump, the temperature being reduced to 120° below zero centigrade—the most intense cold yet obtained."[1]

[1] Researches on Nitrous Oxide; by Geo. J. Zeigler, M. D.

Nitrous oxide may be obtained by dissolving zinc in dilute nitric acid, but when thus prepared the prot-oxide of nitrogen, NO, and the deutoxide of nitrogen, NO_2, are formed; the mixed gas is then allowed to stand over damp zinc or iron filings, when a decomposition of the deutoxide ensues, one equivalent of its oxygen uniting with the zinc or iron, forming the oxide of those metals, the remaining body of gas being reduced to the protoxide of nitrogen or nitrous oxide. When thus prepared, it is necessary that the zinc and acid should be subjected to chemical tests as to their absolute purity, and as most of the commercial zinc and acid contain impurities, the gas is rarely manufactured in this manner. Commercial zinc generally contains traces of sulphur, iron, and arsenic, and nitric acid is usually impure from the presence of hydrochloric and sulphuric acids. Hydrochloric acid may be detected by nitrate of silver and sulphuric acid by a diluted solution of nitrate of baryta.

A much better mode of obtaining nitrous oxide is by decomposing the nitrate of ammonia by heat.

Nitrate of ammonia is a salt, composed chemically of NH_3HO,NO_5; it was formerly called *nitrum flammans*, in consequence of its rapid decomposition when heated to 600°; it deliquesces when exposed to the air, and requires to be kept in well-stopped bottles or jars. Nitrate of ammonia is formed by saturating pure nitric acid with the carbonate of ammonia, then evaporating and crystallizing it. It is prepared in two forms for use—as the crystallized and the fused. The former is the most deliquescent, the latter being preferable particularly for obtaining nitrous oxide. Much of the nitrate of ammonia contains impurities formed either by the use of impure nitric acid, which may have contained hydrochloric acid, or the carbonate of ammonia, which was defective in consequence of the presence of hydrochlorate of ammonia. Under such circumstances the nitrate of ammonia will be impure, and when decomposition of the salt takes place chlorine gas will be liberated and mixed with the nitrous oxide. Chlorine, as an impurity in the salt, can be readily detected, and I would urge each one who prepares nitrous oxide to test for this element in each lot of nitrate of ammonia purchased. The salt is soluble in less than its own weight of water at 60°; and if any chlorine is present, it may be discovered by adding to the solution a small portion of nitrate of silver in solution. The affinity existing between the chlorine and the silver being stronger than for the nitrate, a white curdy precipitate—the chloride of silver—is immediately formed. Or nitrous oxide may be tested by passing it through a solution of nitrate of silver, when the same result would be produced if chlorine be present. When pure nitrate of ammonia is heated, the salt first melts, and boils, the gas being liberated at about 400°, and if care is not taken to properly regulate the heat, the gas may be disengaged so rapidly as seriously to endanger the apparatus in which the decomposition is going on. The nature of the decomposition may be understood by the following diagram :—

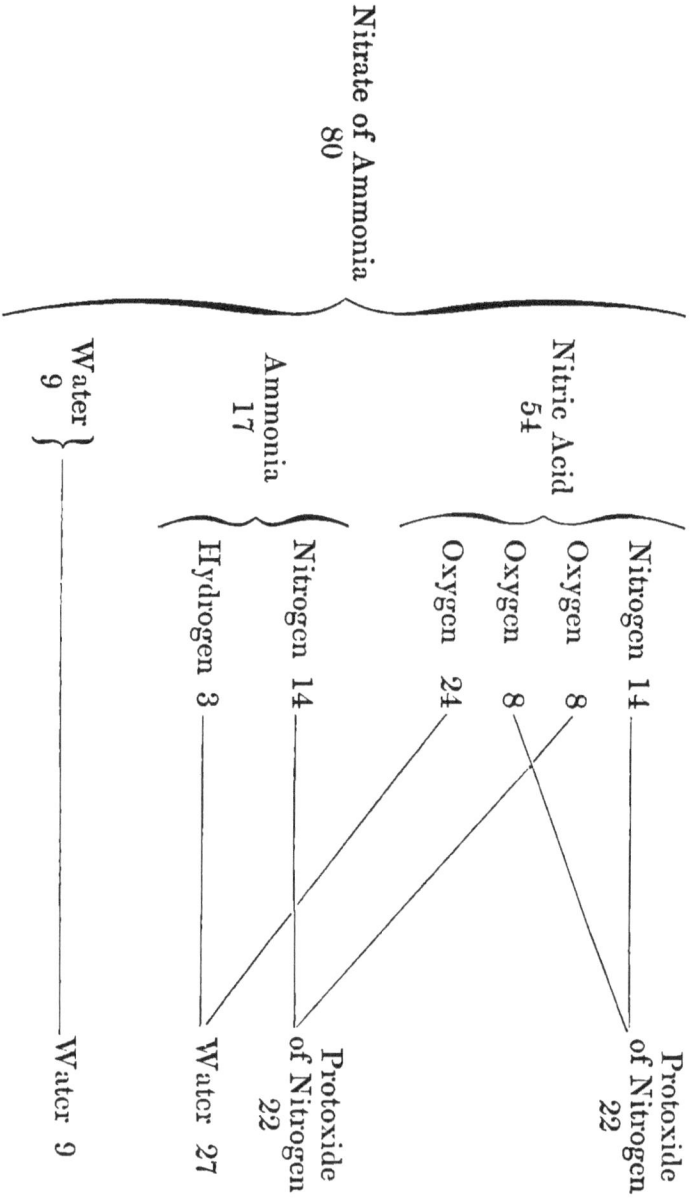

Nitrate of Ammonia
80

Nitric Acid
54

Ammonia
17

Water
9

Nitrogen 14 ———————— Protoxide of Nitrogen 22

Oxygen 8

Oxygen 8

Oxygen 24

Nitrogen 14 ———————— Protoxide of Nitrogen 22

Hydrogen 3

Water 27

Water 9

By the foregoing diagram it will be seen that pure nitrate of ammonia is decomposed into protoxide of nitrogen and water.

When the nitrate of ammonia is heated too rapidly the retort will be clouded by white and red fumes, which by some chemists are considered to be nitrogen and the deutoxide of nitrogen, nitric oxide, or nitrous gas, NO_2. As these gases are liable to be generated even when great care is used, a small quantity of atmospheric air should be present in the gas receiver, as the deutoxide of nitrogen or nitric oxide will take two equivalents of oxygen from the air and become nitrous acid, NO_4, which will be readily absorbed by the water over which the nitrous oxide is allowed to stand. To heat the salt I use a gas stove, Fig. 1, B, manufactured by E. Parrish, of this city, for chemical and pharmaceutical purposes, by which the heat can be easily regulated. The retort should also be shielded by a sand bath, which will furnish a more uniform heat and prevent the fracture of the retort, which is liable to occur when the heat is directly applied to the glass.

II. THE APPARATUS FOR PREPARATION OF NITROUS OXIDE.

The apparatus necessary for generating nitrous oxide is a strong glass retort which should not be

filled more than one-third full of the nitrate of ammonia. The ground glass stopper is not well adapted
for use in the retort, as it is exceedingly difficult to
keep it perfectly tight when the gas is being generated, and if tightened when the retort is heated, it
is with difficulty that it can be loosened when the
retort is cold. I much prefer a good velvet cork,
which entirely obviates the trouble with the glass
stopper. Attached to the retort there should be a glass
bulb, Fig. 1, A, to receive the water which is formed
by the decomposition of the salt. If this receiver is
not made use of, the heated water will collect in the
rubber tubing, and eventually destroy its integrity,
besides to a certain extent impeding the flow of gas
into the receiver.

Nitrous oxide, when generated, requires to be purified, particularly when it is designed for inhalation;
for this purpose it may be washed by passing the gas
through two or three wash bottles, known as Wolfe's
bottles, containing solutions of caustic potash or soda,
sulphate of iron and water, Fig. 2. Saturated solutions of caustic potash and the sulphate of iron should
be made and the Wolfe bottles should be connected
together in the following manner.

The first bottle should have a tube passing from
the retort nearly to the bottom of the bottle, but the
glass tube from the first bottle to the second should

be attached to the stopper and pass nearly to the bottom of the second. The attachment of the second to the third should be in like manner. The object is to have the gas pass directly through each solution.

When thus washed, the gas will be free from chlorine, the binoxide of nitrogen, or other impurity. It may then be passed into a gutta-percha bag, where it may be kept for about forty-eight hours without much loss in quantity or quality, but as endosmotic action takes place between the gas and the atmosphere, it cannot be kept for any great length of time without deterioration, at least such has been my experience.

A simple apparatus for purifying and receiving nitrous oxide may be made by any one in the following manner: Take a good strong barrel—an old linseed oil barrel is very good—attach to the top a supply cock by which it may be filled with water, also a cock from which the gas may be obtained. Let the attachment from the retort pass directly into the barrel nearly to the bottom, fill the barrel with water and apply heat to the retort. When the gas commences to be generated a waste cock at the bottom of the barrel should be opened to allow the water to flow out, giving space in the barrel for the gas; a water gauge should be made use of to show the height of water in the receiver. To obtain the gas from the barrel it is only necessary to turn on the supply of water, and open the supply cock for gas at the top of the barrel. Gas thus prepared will be quite pure as the quantity of water through which it passes will remove the chlorine and the deutoxide of nitrogen, as previously referred to.

A very excellent apparatus in which nitrous oxide may be generated and kept on hand for any length of time, is a metallic gasometer which I have designed and am now using, Fig. 1. It consists of a retort, sand bath, water receiver, purifier, and gas receiver. Two metallic vessels, of best block tin, are made use of, fitting one into the other, the inner cell holding forty gallons. The apparatus is filled with water by the supply cock C; a tube passes directly upwards in the centre of the gasometer, as indicated by the dotted line D; this tube is bent downwards after passing to the top of the water E, the object being that the gas shall pass through and be washed by the water, yet having the tube nearly free from water. The weight of the inner reservoir is also ascertained, and is counterbalanced by the movable weights G. The object of the bent tube and the weights is to offer no obstacle to the uninterrupted passage of gas into the receiver; for, if it has to pass through a column of water, and also raise a metallic cell weighing several pounds, bursting of the retort when the gas is rapidly generated is ex-

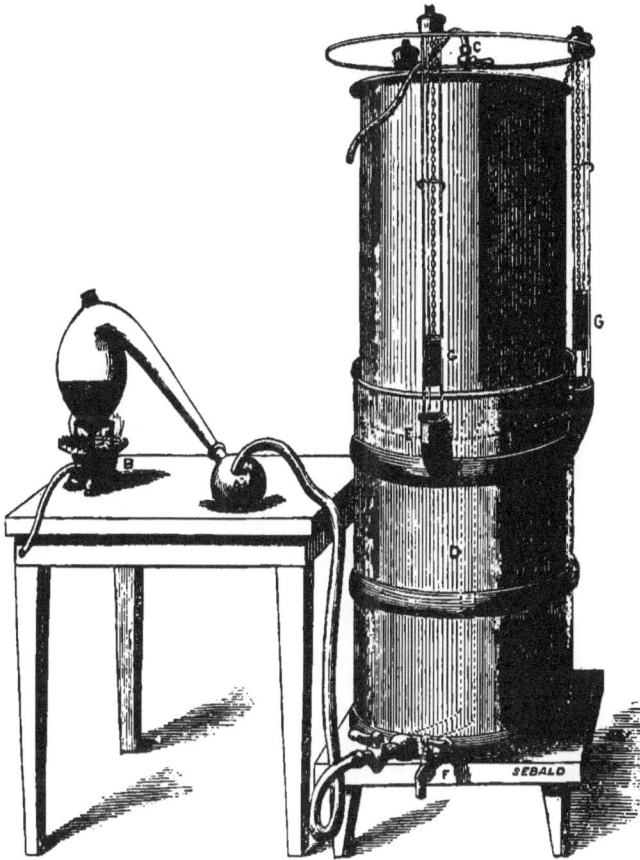

Fig. 1.

The Author's Apparatus for Generating Nitrous Oxide.

Fig. 2.

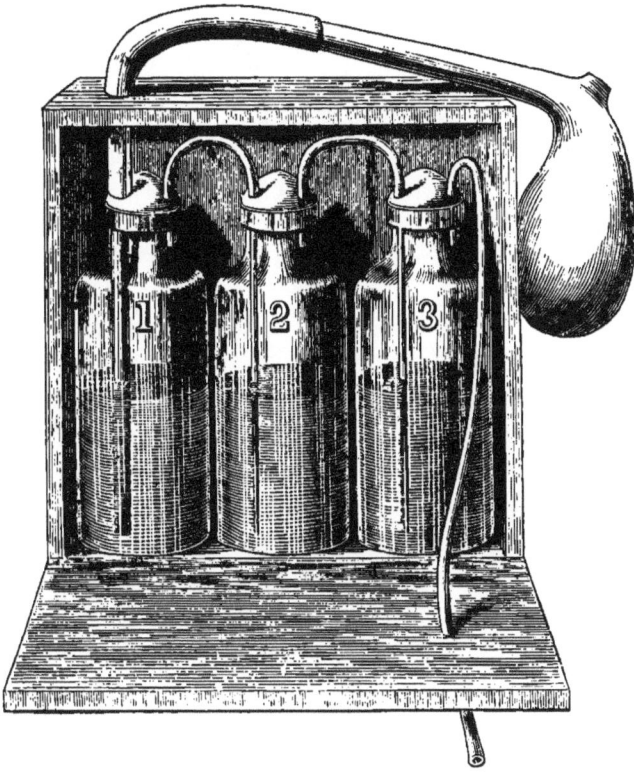

Moseley's Gas Generator.

ceeaingly liable to ensue. When the gas is desired from the receiver it is only necessary to open the supply cock c, and unhook the weights, when it will pass readily into the inhaling bag. The water may be removed at any time by the waste cock F.

Many operators, myself among the number, do not make use of an inhaling bag, but have rubber tubing pass directly from the receiver to the operating-room; to this tubing the mouthpiece or inhaler is attached. This is an excellent method, as the use of the unsightly bag is thus obviated, but it is necessary that a mouth-piece should be used that will not allow the exhala-tions of the patient to pass into the receiver, else the gas is vitiated by admixture with carbonic acid gas.

Much has been said by some whose design is, it would seem to be, to keep this anæsthetic agent from general use, of the necessity of preparing the "gas fresh every day." I would here unhesitatingly state that this is an error, or a wilful deception, as nitrous oxide improves by being kept on hand over water. I have used gas two weeks old with the very best results; it is pleasanter to the smell and sweeter than the fresh gas, being devoid of the pungent odor which can be detected in nitrous oxide when first made.

It may be remarked by some that in an apparatus as just described much loss of gas takes place from the absorption of the gas by the water, the latter

taking up its own bulk. This is true, and when the
water is first introduced, this loss does take place, but
it is only necessary to change the water once per
month, and but one loss takes place, the water ever
afterwards having taken up its own bulk, no subsequent loss will ensue. The advantages of a receiver
of this kind as a purifier over the solutions of potassa
and sulphate of iron in Wolfe's bottles, as previously
described, consist in the fact that these solutions,
when saturated with impurities, will not purify the
gas, and it cannot be readily ascertained what is the
condition of these solutions, but where there is a
great bulk of water there is more certainty that the
noxious gases will be removed. A very excellent
apparatus where the solutions are designed to be used
is the apparatus known as C. H. Moseley's Gas Generator. (Fig. 2.)

Through the kindness of Professor E. Vander
Weyde, I am enabled to describe an exceedingly ingenious and efficient apparatus designed and introduced by him.

The inhaling apparatus, Fig. 3, was first exhibited
in the fall of 1863 before the Dental Association in
New York; since that time it has not been before the
public. I will briefly describe the arrangement, and
then mention some very late and important improve-
ments.

Fig. 3.

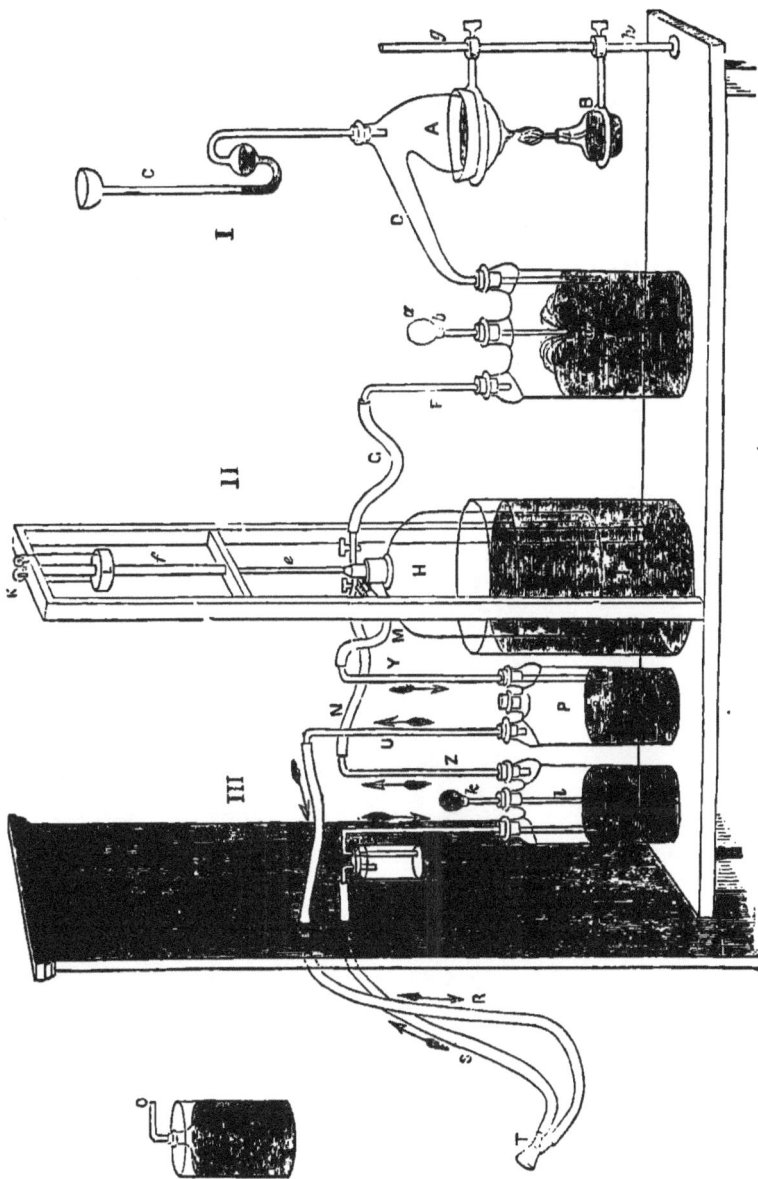

Prof. Vander Weyde's Gas Generator and Purifier.

A bell-jar H is nicely balanced by a weight L over water, in the style of a gasometer, as the substitute for the inhaling bag; it contains the gas, and moves easily up or down during respiration.   However, the chief part is the breathing arrangement, consisting of Wolfe bottles P and L, partially filled with some caustic solution, through which the gas is compelled to go in its passage to and from the mouth of the patient, and which deprives it of the products of respiration, chiefly carbonic acid.   To prevent any exhaled gas from returning to the lungs without passing through the purifying solutions, the mouthpiece T is bifurcated and attached to two flexible tubes, of which R serves to admit gas to the mouth, and S to conduct the exhaled air through the Wolfe bottles back to the reservoir.

To understand this operation, it need only to be observed that the ends of the tubes V and Y are plunged under the surface of the liquid, therefore admitting the passage of the gas only in a downward direction, and also acting as the most perfect valves, preventing all return of gas by the channel by which it came.

In using the apparatus Dr. V. W. found it inconvenient, to prepare small quantities of gas for every patient, the glass reservoirs seldom being large enough for more than one or two doses; he therefore substituted

for the retort A and washing bottle E, a large gasometer of thirty or forty gallons capacity; but afterwards
a small metallic cylinder, strong enough to withstand
a pressure of fifty atmospheres, in which the gas had
been compressed and liquefied at a temperature of
32° Fahr., by a force-pump of a power of thirty-two
atmospheres. By opening a stop-cock part of this gas
is conducted in the reservoir, and a small cylinder may
keep up the necessary supply for a long time.

The advantages of this arrangement are :—

1st. The convenience of having no large reservoirs
or boxes, but a small cylinder of one gallon contents
of liquefied gas is sufficient for one hundred and fifty
to two hundred doses, as it is equivalent to about
eight hundred gallons of gas.

2d. The saving of trouble in making the gas, as the
liquid gas can be transported to the house of the surgeon or dentist, in the same way as the soda water
from the factory.

3d. The gas must necessarily be pure, as the impurities, air, etc., are not condensable at that temperature.

4th. It cannot possibly deteriorate by absorption of
the liquid over which it is kept, or by endosmose of
the bags.

5th. Liquefied gases when expanding become very

cold; the gas must therefore be necessarily very cool when recently expanded.

When the gas is once compressed there is no danger of explosion; this is seen when discharging all the liquid gas from the cylinder, it will then by no means suddenly expand to the gaseous form, but this will take place comparatively slowly under the development of the most intense cold.

In a private letter from Prof. Vander Weyde he assures me that he contemplates erecting a factory for the preparation of liquid nitrous oxide, and will send the liquid in a cylinder to the house of the surgeon. If practicable it will overcome the principal objection to the general use of this anæsthetic.

Nitrous oxide should never be generated in a metallic retort, as it is attended with great danger. I quite recently attempted to prepare the gas in a retort of copper—manufactured expressly for making oxygen gas—and though great care was taken to heat the salt slowly, a violent explosion occurred, fracturing the vessel, and throwing the pieces in every direction. Fortunately, no one was injured, though portions of the retort and gas stove were scattered in every part of the room.   An examination of the pieces of copper demonstrated that a chemical change had taken place, evidence of the ammoniated nitrate of copper being present.   To make the experiment more reliable, and to determine that the explosion did not occur from the choking of the outlets for the gas, or the presence of steam from the reflow of the water, which is set free in the decomposition of the salt— there being no heat under the retort when it exploded —a small quantity of nitrate of ammonia was placed in an open vessel, heat being applied, and while in a liquid condition a thin strip of copper was introduced. Almost immediately the salt assumed a greenish appearance, copious white fumes resembling in smell and causticity nitric acid fumes were given off, followed, in a few moments, by successive explosions. Iron was next tried, but did not produce a like result. That the last named metal may be used with safety is *possible*, but I leave it for others to determine. What was the cause of the explosion I am at a loss to explain, and as each chemist to whom the subject has been mentioned presents some new theory, it yet remains a matter of speculation. The same salt had been repeatedly used in a glass retort without any like result.

III. THE INTRODUCTION OF NITROUS OXIDE AS AN ANÆSTHETIC.

Though discovered in the year 1776, it was not until the year 1800 that much was known of this agent; it then received the attention of Sir Humphrey Davy, who published in his Researches an account of this gas, refuting the idea that had previously been entertained that it was poisonous and irrespirable. That a clear idea of the effect of nitrous oxide was entertained at that time may be seen by a perusal of the following quotation from a paper on nitrous oxide by Dr. T. L. Buckingham.[1]   He says:—

"I quote two cases republished in Brande's 'Manual of Chemistry,' vol. i. pp. 329–30, the first account by Mr. Tobin, the second by Mr. Roget:—

"'On the 29th of April I breathed four quarts from and into a silk bag.   The pleasant feelings produced at first, urged me to continue the inspiration with great eagerness.   These feelings, however, went off towards the end of the experiment, and no other effects followed.   The gas had probably been breathed too long, as it would not support flame.   I then proposed to Mr. Davy to inhale the air by the mouth from one bag, and to expire it from the nose into another.   This method was pursued with less than three quarts, but the effect was so powerful as to oblige me to take in a little common air occasionally. I soon found my nervous system agitated by the highest sensations of pleasure, which are difficult of description; my muscular powers were very much increased, and I went on breathing with great vehe-

[1] Dental Times, vol. ii., No. 1.

mence, not from a difficulty of inspiration, but from
an eager avidity for more air. When the bags were
exhausted and taken from me, I continued breathing
with the same violence; then suddenly starting from
the chair, and vociferating with pleasure, I made
toward those that were present, as I wished they
should participate in my feelings. I struck gently at
Mr. Davy; and a stranger entering the room at the
moment, I made towards him, and gave him several
blows, but more in a spirit of good humor than of
anger. I then ran through different rooms in the
house, and at last returned to the laboratory somewhat
more composed; my spirits continued much elevated
for some hours after the experiment, and I felt no
consequent depression either in the evening or the day
following, but slept as soundly as usual.'

"Dr. Roget states as follows: 'The effect of the
first inspiration of the nitrous oxide was that of
making me vertiginous, and producing a tingling sensation in my hands and feet; as these feelings increased, I seemed to lose the sense of my own weight,
and imagined I was sinking into the ground. I then
felt a drowsiness gradually steal upon me, and a dis-
inclination to motion; even the actions of inspiring
and expiring were not performed without effort; and
it also required some attention of mind to keep my
nostrils closed with my fingers. I was gradually

roused from this torpor by a kind of delirium, which came on so rapidly that the air-bag dropped from my hands. This sensation increased for about a minute after I had ceased to breathe, to a much greater degree than before, and I suddenly lost sight of all the objects around me, they being apparently obscured by clouds, in which were many luminous points, similar to what is often experienced on rising suddenly and stretching out the arms, after sitting long in one position. I felt myself totally incapable of speaking, and for some time lost all consciousness of where I was, or who was near me. My whole frame felt as if violently agitated; I thought I panted violently; my heart seemed to palpitate, and every artery to throb with violence; I felt a singing in my ears; all the vital motions seemed to be irresistibly hurried on, as if their equilibrium had been destroyed, and everything was running headlong into confusion. My ideas succeeded one another with extreme rapidity, thoughts rushed like a torrent through my mind, as if their velocity had been suddenly accelerated by the bursting a barrier which had before retained them in their natural and equable course. This state of extreme hurry, agitation, and tumult, was but transient. Every unnatural sensation gradually subsided; and in about a quarter of an hour after I had ceased to breathe the gas, I was nearly in the same state in which I had been at the

commencement of the experiment. I cannot remember
that I experienced the least pleasure from any of these
sensations.' "

At this time Sir Humphrey Davy made use of the
prophetic language, which has subsequently been so
completely verified: " As nitrous oxide in its exten-
sive operation seems capable of destroying physical
pain, it may probably be used with advantage during
surgical operations in which no great effusion of blood
takes place."

For forty-four years this prophecy was unfulfilled,
no use being made of nitrous oxide, except as an exhilarant in the lecture-room. At one of these public
lectures in 1844, Horace Wells, of Hartford, Connec-
ticut, on witnessing the peculiar physiological effects
conceived the idea of using this agent for the purpose
of obtunding sensibility, and besides having a tooth
extracted himself, also extracted others for different
individuals; but being unaware of the proper mode
of administering it, and being deficient as an experi-
mentalist, he failed to secure uniform results, and
became discouraged with it. The credit of demon-
strating the practicability and possibility of inducing
anæsthesia in surgical operations belongs to Dr. Wells,
and though he cannot derive the benefit, yet a grate-
ful nation should attest the value of this great boon
by donating to his family such a testimonial as will,

in a slight manner, at least, compensate them for this priceless contribution to science, and the relief of suffering humanity.

Into what insignificance do the narcotic agents—as mandragora, morion, opium, or Indian hemp, used for obtunding pain, sink beside this discovery, and from which have sprung as offshoots from this parent stock, ether and chloroform. Well does the ancient poet refer to Helen, who tempers with drugs the draught she offers to Mellenius and his friend :—

"To clear the cloudy front of wrinkled care,
    And dry the tearful sluices of despair,
    Charmed with that virtuous draught the exalted mind,
    All sense of woe delivers to the wind."

But in this age no noxious drugs are necessary, for in a moment's time anæsthesia may be induced and in as short a period perfect restoration to consciousness take place; that this is so is due to the practical investigation of Dr. Wells, and as a dentist I feel that had our profession never have contributed anything else but nitrous oxide by Dr. Wells and ether by Dr. Morton, it would deserve lasting honor from mankind.

IV. MODE OF ADMINISTRATION.

Nitrous oxide for medicinal purposes is often associated with liquids or vapors, but as my experience

with it is simply as an anæsthetic, I shall confine myself to its consideration under that head. For the
purpose of an exhilarant it is only necessary that a
person should inhale a small quantity, say three or
four gallons of the gas; the agent being removed before
anæsthesia is induced, the admixture of atmospheric
air which takes place produces in many persons a decidedly stimulating effect, not developing, however, as
is usually believed, the propensities of the individual.
My own experience has been directly the reverse, for
I have usually found that those who were ordinarily
the most quiet and retiring in disposition became
under its influence the most boisterous and unruly.
I shall, however, refer to this subject when speaking
of its properties.

For the production of anæsthesia it is necessary
that nitrous oxide should be administered unmixed
with atmospheric air. For this purpose a suitable
inhaler should be made use of; these may be obtained either of vulcanized rubber, Fig. 5, or of
silver. I have until recently used an inhaler shown
in Fig. 4. It consists of a silver mouth-piece, which
covers the lips, the patient being directed to close the
mouth upon the hard rubber tube A. This tube is
designed, in case of forcible closure of the mouth, to
slide from the silver inhaler, and being fixed between
the teeth, the mouth is opened sufficiently for the in-

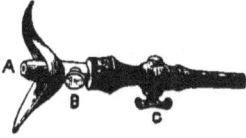

Fig. 4.

Fig. 5.

Fig. 6.

Fig. 7.

(37)

troduction of the forceps.   It has also a saliva cup B, which is designed to catch the saliva which flows with many persons very copiously while the gas is being administered.   The stopcock C arrests the flow of gas when desired.   This inhaler is a modification of one invented by C. Bullock, of Cambridgeport, Mass.   His inhaler is certainly very ingenious, but being quite complicated has not produced satisfactory results in my hands.   It is designed to supply the gas for inspiration and not allow the exhalations from the person to pass back into the gas-bag.   The one I have referred to, Fig. 3, is deficient in this respect, as it allows the carbonic acid gas given off by the patient to become mixed with the nitrous oxide.   A good inhaler that will obviate this difficulty is much needed, and my attention has recently been called to one invented by Dr. D. H. Goodwille, of New York city. (Fig. 7.)

Its object is to furnish nitrous oxide to the patient unmixed with carbonic acid gas.   To accomplish this two rubber valves, A and E, are made use of in the stopcock, one opening inwards only to allow the gas to pass in at an inspiration while it is closed by the force of the exhalation which opens the second or waste valve E.   It is also designed by the inventor to collect and purify the exhaled gases by attaching a rubber tube and bag into which the waste valve will open. It has also a hard rubber mouth-piece, C, which covers the nose as well as the mouth. It is an exceedingly ingenious and effective appliance.

In all cases where nitrous oxide is administered, particularly if an inhaler is used that allows the person to breathe from and into the bag, it is necessary to instruct the patient as to the mode of inhalation previously to the time of operating. This is by no means an easy matter, as I have repeatedly found when administering the gas for surgical purposes at the different hospitals and surgical clinics of this city. If the operation is only one requiring a minute or two for its performance, a bag holding not less than seven or eight gallons of gas should be used, but if a longer time is required, the gas-bag will necessarily require to be of greater capacity. The patient should then be instructed to take full inspirations from the bag, previously, however, to the first inhalation being required to empty as perfectly as possible the lungs of atmospheric air. The nostrils should be tightly closed by the person who is administering the gas so that no atmospheric air passes in with the nitrous oxide. An assistant should always be present to hold the bag or adjust the rubber tubing so that the flow of gas is not arrested.

As a proof of the necessity of this caution I would refer to an article by Dr. W. P. Moon, in the " Medical and Surgical Reporter." " In witnessing the exhibition

of anæsthesia by nitrous oxide gas on one occasion at
the Pennsylvania Hospital, I think a patient was as
nearly dead as I should care to see any one, and have
my doubts if it is without its dangers either through
unskilful management or improperly prepared gas."
In this case referred to I have been assured by the
surgeons operating that the difficulty arose simply
from the tubing from the gas-bag becoming coiled so
as to cut off the supply of nitrous oxide.   Anæsthesia
will usually occur in from half a minute to two
minutes, and if it is not induced in three or four
minutes the eight gallon bag should be removed and
one much larger substituted.   This I consider impera-
tively necessary, as carbonic acid is given off in greatly
increased quantity while nitrous oxide is present in
the system.   A simple experiment can be tried with
gas that has been inhaled for two or three minutes;
by placing in it a lighted taper, it will burn but feebly
if at all, not supporting combustion like pure nitrous
oxide in an active manner, as will be referred to
hereafter.

The first evidence of anæsthesia with the majority
of persons is snoring, not the deep stertorous breathing
as seen in anæsthesia from chloroform or ether, but
more like the snoring of deep, heavy sleepers.   Snoring
does not usually occur when the patient is in a recum-

bent position, the only test of insensibility being the relaxed condition of the muscular system.

When the operation is a protracted one, the inhaler should every few moments be removed and the lungs well filled with atmospheric air, when the gas may be again administered without any interruption of the anæsthetic condition. In this manner a patient may be kept for a long time under its influence without difficulty or danger. I would here remark that the longest time that I have kept a patient under its influence was about twenty minutes, the operation being the removal of a scirrhous breast by Prof. Joseph Pancoast. I have several times kept patients ten or twelve minutes under its influence without any unpleasant result.

Prof. Carnochan, in the "Medical and Surgical Reporter," February 10, 1866, says: "Since my letter in December I have performed four more capital operations on adults, viz., one amputation of the thigh, one of the leg, the removal of a tumor from the side, and the extraction of a cataract, making in all, since last July, seven successful capital operations under the influence of anæsthesia produced by the nitrous oxide gas. I have also during this time used chloroform and ether in many operations, and my opinion in regard to the superiority of the nitrous oxide as an anæsthetic is still unchanged. I believe, however, that there is great room for improvement in the mode of administration of the gas; one principal fault at present being the repeated inhalation of the same material. An instrument which will act by valvular arrangement, as in Reed's stomach pump, would obviate this difficulty, and I have no doubt but that some skilful mechanism will produce one that will meet the necessary requirements."

The inhaler which I have recently designed and which I now use answers the requirements demanded. (Fig. 6.) The position of the valves is the same as in the inhaler invented by Dr. D. H. Goodwille, but it possesses these advantages. First, the cap A, which covers the mouth and nose of the patient is made of flexible rubber, and as a consequence when the gas is inhaled the appliance is drawn on to the face, no matter what may be its conformation or the age of the person, making an air-tight mouth-piece. Second, the valves C and E in the stopcock close with a spring like the keys of a flute. These springs overcome the pressure of gas, and do not allow the supply valve to be constantly open, which is the fault in most of the inhalers, there being a constant waste of gas. The hard rubber tube D is arranged to slide from the mouth-piece in case of forcible closure of the mouth. A rubber tube F with gas-bag is attached to the mouth-piece at the point indicated by dotted lines, into which the waste valve opens. The nitrous oxide which has escaped or has been exhaled with carbonic acid gas is collected in the bag, and this impure gas may be thoroughly prepared for future inhalation by passing it through a solution of lime-water or caustic potash; the latter is rather preferable. The affinity which exists in either of these solutions for carbonic acid is exceedingly strong, and with but little trouble the gas may be preserved and purified. A single Wolfe's bottle may be used in the following manner: To the tube, which passes nearly to the bottom of the bottle, attach the bag of impure nitrous oxide; to the glass tube which commences above the solution near the cork a rubber tube should pass to the regular gas receiver. By simply making pressure on the bag of impure gas it will pass through the solution into the receiver, it having been deprived of its carbonic acid gas. Since the introduction of this inhaler I have abandoned the use of the gas-bag, which is so exceedingly unsightly, and now have a rubber tube passing from the supply valve c, Fig. 1, through the partition into the extracting-room; to this the inhaler is attached.

Nitrous oxide, when administered unmixed with atmospheric air induces perfect repose, there being an entire absence of muscular contractions so common with other anæsthetics. This constitutes one of its advantages; another consists in the fact that it does

not induce emesis, which frequently follows the exhibition of ether or chloroform.

V. CHEMICAL PROPERTIES OF NITROUS OXIDE.

As previously remarked, nitrous oxide is composed
of the same elements as atmospheric air, with the exception that the former contains *one-third* of oxygen
to *two-thirds* of nitrogen, while the latter contains *one-
fifth* of oxygen to *four-fifths* of nitrogen. Nitrous
oxide is therefore richer in oxygen than atmospheric
air. This fact has been taken advantage of by persons
unskilled in chemical knowledge to advocate the use
of nitrous oxide on this ground; but this position is
not tenable, as, if this were admissible, nitric acid,
NO_5, could be urged as still better from its containing
more oxygen than either, when it is well known that
it is very poisonous, and will not support life even for
a moment. Nitrous oxide differs from atmospheric
air in the fact that in the former the elements are in
chemical combination, while in the latter they are in
the form of a mixture. A proof that nitrous oxide is
a chemical combination of the two elements may be
instanced by the fact that it has both a taste and smell,
while air, a simple mixture, possesses neither. As an
example of the difference between a chemical combination and a simple mixture of elements I might cite

hydrogen and oxygen; when these gases are united as a mixture, they may be inhaled for a limited period without bad results; but when chemically combined, as in water, they will not support life for a moment. A better reason for the harmlessness of nitrous oxide consists in the fact that it is, like oxygen, an energetic supporter of combustion, and hence a supporter of life. The following experiments are detailed as an evidence of its power of supporting combustion. A piece of wood may be lighted in the open air; if the flame is blown out, and, while lighted, it is plunged into a jar of nitrous oxide, it will be immediately kindled into a flame again. This experiment may be repeated several times with the same result, until the contents of the jar have become mixed with atmospheric air and carbonic acid gas. By some writers on chemistry this experiment is said to have been unsuccessful, but after repeated trials I have always met with a uniform result. Sulphur, which burns feebly in the open air, will, when plunged into nitrous oxide, burn with great brilliancy and beauty; the same result takes place with phosphorus, but both require to be heated before they are introduced into the gas. Sulphide of carbon or ether introduced into this gas on a lock of cotton burns with great splendor. There is also present a peculiar halo of a reddish color, supposed by Brande to arise " probably from the combustion or incandescence of nitrogen at a high temperature."

Were it not that nitrous oxide thus supports combustion, its administration in the manner directed would be attended with fatal results, but as it promotes chemical combustion in the system, it may be administered in this way without danger. It differs from oxygen in its great solubility in water; it is also not as efficient a supporter of combustion as oxygen, as iron wire will not burn in it.

VI. PHYSIOLOGICAL PROPERTIES OF NITROUS OXIDE.

Respecting the *modus operandi* of protoxide of nitrogen much difference of opinion exists; some contending that it acts upon the system by means of its oxygen or nitrogen singly, others assuming that it produces its effect by and through the medium of these elements combined. At this stage of use it would seem as if it were but proper to wait for a more ex tended experience, when it is hoped that more light may be given on this particular branch of the subject. Dr. Ziegler, whose experience with it as a medicinal agent extends over a period of seventeen years, says: " Protoxide of nitrogen is thus altogether unique in being a rapidly diffusible, potent, general, and per-

manent stimulant, intensifying all the vital actions,
supplying elements of nutrition, exerting both a
material and dynamic influence upon the animal
economy, and in having a direct physiological compa-
tability therewith both functional and organic. Hence
with respect to the predominant characteristics of protoxide of nitrogen it is truly *sui generis*, though closely
allied in chemical constitution, material properties, and
sanitive effects with atmospheric air, which may be
regarded in fact as its natural prototype, differing
thereform apparently more in the proportion of its con-
stitutional elements—nitrogen and oxygen—and in
the manner of their association, than in any other
essential respect, although its action upon the vital
economy is more energetic, definite, and perceptible
than that of the latter, varying more in the degree,
perhaps, than in the nature of its physiological influence. Still, though there is thus a very strong
analogy between protoxide of nitrogen and atmospheric
air, yet in their relative effects upon the animal organ-
ism they vary somewhat materially, as there is a
striking difference in the ratio or measure as well as
in the kind of action thereupon, for the biological
effects of the former are not only more intense, con-
centrated, and manifest, but also more decidedly stimulant than those of the latter, though they are like-

wise of a general, very bland, and highly invigorating character."[1]

Nitrous oxide, like alcohol, acts as a stimulant to the system. In moderate quantity both act as exhilarants, producing intoxication. When taken in large doses, both act by inducing narcotism and insensibility. Nitrous oxide differs from ether and chloroform in its power of supporting combustion and respiration; the two latter being deficient in this respect probably act as sedatives by depressing the nervous system and thereby reducing vital action below its normal standard. Nitrous oxide, on the contrary, by its stimulating effect increases nerve force and brings vital action above its standard; the effect of both, however, is the same, viz., total insensibility. Whether it is more dangerous to elevate or depress the human economy time must determine, but my own impression is that the former is much less hurtful.

It has been supposed by some writers on this subject that nitrous oxide when administered for the purpose of anæsthesia, produces its peculiar effects by over stimulation in the same manner that oxygen acts upon an inferior animal by turning all the blood of the system into arterial, no change taking place in the capillary system. This view I believe to be erroneous,

[1] Researches on Nitrous Oxide, page 14.

and from certain reasons evident to all. First, the blood which flows at the commencement of an operation is very apt to be dark in appearance, showing the effect of carbonic acid gas. Second, the dark and livid color of the lips tells us unmistakably that carbonic acid is present in the blood. My opinion is that the chief danger in nitrous oxide consists in the rapid formation of carbonic acid gas, which is due to the excessive disintegration and waste that continue while the system is under its influence. This carbonic acid gas is not all expelled from the lungs at the period of expiration, but, as it is a heavy gas, a portion remains in the lungs, occupying the space of the residuary air which is normally present; this and the super oxidation going on in the capillary system would require us ever to be on the guard for carbonization of the blood. It in my judgment is another proof of the necessity of an inhaler which will allow no exhalations to deteriorate the nitrous oxide. If a mouse is placed under a receiver and allowed to breathe nothing but nitrous oxide, it will die, and on opening the heart and bloodvessels the blood will be found to present the dark color usually present when death results from asphyxia. This experiment would seem to prove that the agent requires care and judgment in its administration.

The influence of nitrous oxide on the circulation

in the great majority of instances is not marked; generally the action of the heart is somewhat accelerated, but in some few cases a rapid change takes place, the pulse being elevated very rapidly, requiring that great care should be used, and indeed in every case, as with all anæsthetics, the pulse should be carefully watched through the operation and at the time of administration. Respiration becomes somewhat labored if the nitrous oxide is too long inhaled; therefore, as a rule, the inhaler should be removed from the mouth in a few seconds after insensibility has been induced; it may in a few seconds be again applied without interference with the anæsthesia. The influence of nitrous oxide on the organs of generation may be marked when it is used as a medicine combined with some liquid, as water; but I have never observed any aphrodisiac effects, as referred to by Dr. Ziegler, when administered for anæsthetic purposes. It does, however, have a special tendency to increase the urinary secretion; children while under its influence are quite apt to evacuate the contents of the bladder, and adults usually feel an irresistible desire to void water immediately after its exhibition.

The anæsthetic properties of protoxide of nitrogen are due to stimulation, and in this respect it differs from ether and chloroform, which act by producing sedation. Neither of these last-named agents will support combustion or life, unless mixed with atmospheric air. Though nitrous oxide acts as a stimulant, it seems to set aside the law which usually governs stimulants referred to by Dr. Geo. B. Wood in his work on Pharmacology and Therapeutics. He says: " One of the laws of all stimulation, whatever may be its degree, is that it is followed by a depression proportionate, at least approximately to the previous exaltation of the function or functions excited;" and another law quite as universal is that the more rapidly any stimulant effect is induced the greater will be the subsequent depression. In this respect the action of nitrous oxide is quite unique, as patients rarely state that they feel any subsequent depression. Another advantage in the use of nitrous oxide consists in the fact that it does not have any tendency to the production of emesis unless it is administered *immediately after eating ;* this is a decided advantage, as all must be aware who have used to any extent ether or chloroform. In my experience with nitrous oxide as an anæsthetic I have found that its influence upon the system usually passes off entirely in about three or four minutes after the removal of the inhaler ; but there are few who are insensible to pain for this length of time; the period of insensibility is usually not more than a minute or a minute and a half. Patients after its administration generally express themselves as feeling quite well, *not* particularly invigorated, as has been urged by some enthusiasts on the subject, but as well as when they commenced its inhalation. There is one more peculiarity of nitrous oxide, which to dentists is of infinite value. All know with what feelings patients commence the inhalation of an anæsthetic. If ether or chloroform is used and the patient awakens from the anæsthetic state, it is with difficulty that he can be induced to again inhale the agent, but with nitrous oxide the sensations experienced are usually so agreeable that a feeling of confidence is engendered, and it is extremely rare to find any one who objects to its re-administration.

VII. CONDITIONS IN WHICH NITROUS OXIDE IS CONTRA-INDICATED.

Protoxide of nitrogen it may justly be urged is less dangerous than either ether or chloroform, for the reasons, as previously stated, that it is a supporter of life; but, like all other valuable medicinal and anæsthetic agents, it may produce dangerous and fatal results by inducing great waste of vital force or modifying the conditions of the fluids of the body. In a case which recently came into my hands I am satisfied that death would have resulted in a few moments if the inhalation had not been quickly discontinued and the patient restored by the use of diffusible stimulants.

The case was an exceptional one, and it is detailed for the purpose of cautioning as to the indiscriminate use of nitrous oxide. The lady was one of that noble band of women who sacrificed health and comfort during the recent four years of war to minister to our brave heroes on the battle field, in the camp, and the hospital. She was with the grand army of the Potomac from the first battle of Bull Run to the final surrender of Lee's army. While on the battle field she experienced a sunstroke, from which she nearly lost her life. Since that period there has existed a tendency to cerebral inflammation; this fact being known, I administered nitrous oxide with the greatest caution, a twelve gallon bag being used; after three or four inspirations she gave evidence of the most severe pain of the head and over the region of the heart; the pulse ran up to one hundred and fifty, became feeble and intermittent, the breathing labored, while the countenance quickly assumed a livid and death-like aspect. There was every appearance of speedy dissolution, and I doubt not that death would have resulted in a few moments if the anæsthetic had not been removed. She did not recover from the effects during the whole day, but is at present quite well. She stated to me subsequently that a predisposition to functional derangement of the heart existed in her family, but that it had not exhibited itself with her. There are other conditions where I consider nitrous oxide contraindicated as an anæsthetic—in cases of seriously diseased heart, either functional or organic, in active congestion, or acute inflammation of the brain, lungs, or kidneys.   I should use it with the greatest caution where there was a general plethoric condition or where there was a tendency to the hæmorrhagic diathesis.   Subsequent experience will doubtlessly prove that there are other conditions in which it is contraindicated, and in some conditions that have been referred to, time may prove its use to be admissible. The danger of nitrous oxide consists in the fact that it increases oxidation of the fluids and solids of the body, and also acts as a stimulant to the brain and nervous system; where there exists any predisposition to congestion or inflammation, the administration of nitrous oxide may develop this latent tendency, and a fatal result ensue.   Its advantage over other anæs-thetics, even in the morbid conditions referred to, consists in the fact that being so rapidly absorbed any bad effects can be quickly detected, while its rapid elimination from the system favors a speedy recovery. My object in referring especially to these points is to combat an idea which is being promulgated to the public in lectures and advertisements, that it " can be given in all sorts and stages of disease," and because its chemical composition is oxygen and nitrogen, " it only makes people live faster."

VIII. ALLEGED DEATHS FROM INHALATION OF NITROUS OXIDE.

One of the most marked instances of a fatal result following the inhalation of nitrous oxide is the following, and I quote the case in the language of the operator from the *Dental Times*, vol. i. page 157, by José R. Brunet, D. D. S.

" On the 14th of January, 1864, Mr. Samuel P. Sears called at my office for the purpose of having two right molar teeth extracted. He asked to have the 'Laughing Gas' administered, and I proceeded so to do, in the same manner as for any other patient. His general appearance was good, and he told me he had taken chloroform, but did not state at what time or by whom it had been administered, as I was very busy at the time. The teeth were extracted at $4\frac{1}{2}$ P. M., he being placed under the influence of the gas, but not thoroughly. He did not move during the time of extraction, and he appeared to recover in about five minutes, and as I did not observe any unusual symptoms during or subsequent to the operation, I left him in the chair with an assistant, and went to an adjoining room and administered the same gas to a lady. After so doing, went back to Mr. Sears, who

told me he felt sick, and that he had been taken with
an attack of the diarrhœa, also expressing a desire for
fresh air.   Perceiving that he appeared to labor under
some difficulty in respiration, Dr. Dane was immediately sent for, who examined him and found his
lungs greatly congested.   I went and notified his
parents, and when I returned he was dead.

"A post-mortem examination was made next day,
at twelve o'clock, by Dr. George B. Bouton, in the
presence of Dr. Dane and others.   Both lungs were
found bound by old pleuritic adhesions of an exceedingly firm character, the right much more than the
left, which was about three-fourths covered.   The
only portion of the lung tissue which seemed to be
available for the purpose of oxygenation was the lower
half of the right; all of the rest was so covered with
tubercular deposition, patches of hepatization and
vomica as to seem comparatively useless.   There were
also six cavities in the left lung, each of which would
contain an average of half a fluidounce.   There was
also a mass in the lobe of this lung of an almost car-
tilaginous consistency, of the bulk of about three
ounces, made up apparently of tubercular depositions.
All the available portions of the lungs were found
greatly congested; a portion of the apex of the right
lung was free from blood, its tissue being so changed
as not to admit of engorgement of blood or the per-

meation of air.   There were no well-marked changes in any of the other organs examined, except in the right kidney, a drop of pus being noticed in the pelvis.

" Death occurred from congestion of the lungs, occasioned by the nitrous oxide.   The gas was pure, having been administered previously to and after the accident to different parties.   I have also administered to hundreds of patients both ether and chloroform, separately and combined; have also exhibited the nitrous oxide ever since it has been used as an anæs- thetic for the purpose of extracting teeth, and though some have exhibited unpleasant symptoms, have never had them, except in this instance to be attended with fatal results.   I have understood from a member of the family of Mr. Sears, that his physician thought it doubtful whether he would last through the winter, as his lungs were so greatly diseased.

" I now recommend a thorough examination of every patient to be made before administering the gas, its effects being, I believe, where the system is diseased, the same as that of any other anæsthetic."

NEW YORK, January 15, 1864.

From an examination of the evidence in this case, it would seem as if death was the result of pulmonary congestion, the patient being in a condition which favored that disease, and it was doubtless developed at that time by the injudicious administration of nitrous oxide. Another case is reported in the "Dental Cosmos," vol. v. page 491, by G. Q. Colton: "Dr. Gillman informs me that Miss Bell, of Vermont, inhaled a small dose of the gas for sport (not for anæsthetic purposes), with several others on Friday afternoon January 29th. Came out of it as well as any one ever does. Attended a party the same evening, as well to all appearance as ever, full of life and frolic; was taken sick on the next day (Saturday) and died on Wednesday."

"These facts are fully corroborated by the 'St. Alban's Messenger,' which calls her disease inflammation of the meninges of the brain and spinal cord." Dr. Ziegler remarks that this doubtless was the disease, "yet it is quite probable that the excitant influence of nitrous oxide was injurious in so far as it promoted the disorder of the brain and spinal marrow, while the fulness of 'life and frolic,' so freely manifested after its use, was both an immediate effect of its stimulus, and a precursor of the fatal affection, it being a well-known fact that increased functional activity is a primary and constant concomitant of inflammation. Hence it is more likely that, in this case, also nitrous oxide was accessory to the death; even if it did no more than encourage an inflammatory tendency, and thus partially act as an exciting cause of derangement, the predisposition previously existing, of course to such a degree as to render the extraneous stimulus from this or any other agent absolutely dangerous."

IX. CONCLUSION.

When it is known with what recklessness this gas is used, it is only surprising that so few deaths are reported. In conversation with several dentists on this subject, I have been repeatedly assured that the gas that remained in the bag after inhalation, they always replaced in the receiver; and many dentists, I am informed, have a large bag, from which all the patients of the day inhale, and into which they exhale. That this occurs, is, in many instances, due to the limited information which has existed as to the properties and administration of nitrous oxide as an anæsthetic. It is to guard against overweening confidence, and to point out the advantages and disadvantages in as plain and practical a manner as possible, that I have undertaken to present a work on this subject, feeling fully aware that time and use will but demonstrate its value and usefulness, if it be judiciously and carefully administered. I feel also assured that a disregard of these cautions cannot but be attended with danger, or probably fatal results, while an anæsthetic of intrinsic value must necessarily from

such negligence be brought into disrepute. With regard to its *modus operandi* and physiological effects,
I know that much remains to be learned, and it is
sincerely hoped that scientific investigators will consider this a not unworthy field of research.

Through the kindness of many friends, practitioners
of medicine and surgery, I have been enabled to test
the adaptability of nitrous oxide to surgical practice, .
having administered it in a large number of capital
operations. Each of these gentlemen have uniformly
expressed themselves gratified with the result. These
operations I have thought it best not to detail, as
many of these surgeons have given nitrous oxide
but a limited trial, and, doubtless, would not, as yet,
feel prepared to base evidence in its favor until after
further experience. Its introduction into two of the
most prominent clinical institutions in the United
States, viz., the Wills Ophthalmic and Philadelphia
Hospitals, indicates that its advantages are being
appreciated for clinical practice where time is a consideration, and I doubt not that further experiments
will but demonstrate that new uses will be made of
this agent, as valuable perhaps as its anæsthetic
properties.

PRICE LIST.

DR. BARKER'S APPARATUS, Complete (Fig. 1, page 19), including Gasometer (40 gallons), Condensing Chamber, Retort (½ gallon), Sand Bath, Gas Stove, Rubber Tube, large (6 feet), Rubber Tube, small (8 feet), Inhaling Bag (7 gallons), Rubber Mouthpiece (Fig. 5, page 37), and five pounds fused Ammonia $80 00

DR. MOSELEY'S APPARATUS, Complete (Fig. 2, page 21), including Generator, Two sets Chemicals, Retort (½ gallon), Sand Bath, Gas Stove, Rubber Tube, small (6 feet), Rubber Gasometer (40 gallons), Inhaling Bag (7 gallons), Rubber Mouthpiece (Fig. 5, page 37), and five pounds fused Ammonia 54 00

DR. BARKER'S GASOMETER (40 gallons), (Fig. 1, page 19) 55 00

DR. MOSELEY'S GENERATOR—Retort, Inhaling Bag (7 gallons), Rubber Mouthpiece, and Two sets Chemicals . . 25 00

DR. MOSELEY'S GENERATOR—Retort and Two sets Chemicals 15 00

Rubber Inhaling Bags, 5 gallons	6 75
" " 6 "	7 25
" " 7 "	7 75
" Gasometer, 40 "	18 00
" Tube, large—per foot	60
" " small "	35
" Mouthpiece (Fig. 5, page 37)	2 00
" " Dr. Barker's (Fig. 6, page 37) . .	6 00
" " Dr. Goodwillie's (Fig. 7, page 37), with two hoods	10 00
Gas Stove (Fig. 1, page 19)	5 00
Bohemian Glass Retorts, tubulated (½ gallon) . . .	1 50
Condensing Chambers, glass	75
Pan for Sand Bath	30
Ammonia, best fused, per pound	90
" " crystallized, per pound	75

Boxing extra.

For sale by Rubencame & Stockton, 825 Arch Street, Philadelphia, Pa.

www.ingramcontent.com/pod-product-compliance
Lightning Source LLC
Chambersburg PA
CBHW021632270326
41931CB00008B/982